Electricity

Copyright © by Harcourt, Inc.

All rights reserved. No part of this publication may be reproduced or transmitted in any form or by any means, electronic or mechanical, including photocopy, recording, or any information storage and retrieval system, without permission in writing from the publisher.

Requests for permission to make copies of any part of the work should be addressed to School Permissions and Copyrights, Harcourt, Inc., 6277 Sea Harbor Drive, Orlando, FL 32887-6777. Fax: 407-345-2418.

HARCOURT and the Harcourt Logo are registered trademarks of Harcourt, Inc., registered in the United States of America and/or other jurisdictions.

Printed in the United States of America

ISBN 978-0-15-362068-3
ISBN 0-15-362068-4

1 2 3 4 5 6 7 8 9 10   179   16 15 14 13 12 11 10 09 08 07

Visit *The Learning Site!*
www.harcourtschool.com

Lesson 1

How Are Electricity and Magnetism Related?

VOCABULARY
electricity
electromagnet

Electricity is a form of energy. It is produced by moving electrons. Look at the fountain. The lights on the fountain use electricity.

An **electromagnet** is a magnet that runs on an electric current.

READING FOCUS SKILL
MAIN IDEA AND DETAILS

The main idea is what the text is mostly about. Details are pieces of information about the main idea.

Look for information about how electricity and magnetism are related and details about each way they are related.

Electricity

In an atom, electrons move around the nucleus. Electrons can also move from one atom to another. This movement of electrons produces **electricity**. Electricity can be changed into other forms of energy. Electricity can be used to produce light energy and sound energy. It can produce mechanical energy and heat energy.

 What movement produces electricity?

The heater uses electricity to produce heat energy. Heat energy then warms a room. ▼

Electricity and Magnetism

An **electromagnet** is a magnet that uses electricity. It is made by coiling wire around a piece of iron. An electric current runs through the wire. A single piece of wire does not make a strong magnet. More wire is coiled around a piece of iron to make the magnet stronger.

Electromagnets are different from regular magnets. They can be turned on and off. Electromagnets are used to lift heavy iron or steel objects. The electromagnet is turned on to pick up the object and move it. Then the electromagnet is turned off to put the object down.

◀ The crane has an electromagnet that is turned on to pick up metal.

Electromagnets are also used in smaller machines. Doorbells use electromagnets. When you press the button, electricity moves through the electromagnet. The magnet pulls a steel spring. The spring moves a clapper, which makes the ringing sound.

Magnets can also produce electricity. In an energy station, a coil of wire turns inside a magnet. This produces electricity. The electricity then travels across other wires so it can be used in homes, schools, and offices.

 How can an electromagnet be made stronger?

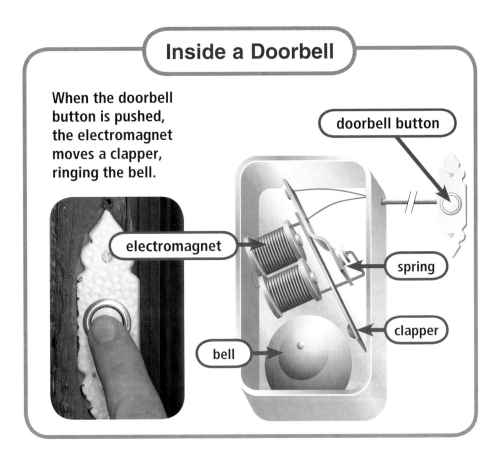

Inside a Doorbell

When the doorbell button is pushed, the electromagnet moves a clapper, ringing the bell.

electromagnet

doorbell button

spring

clapper

bell

Electric Motors

Have you ever stood in front of a fan on a hot day? Electricity makes the fan move. The motion of the fan is produced by an electric motor.

An electric motor has a coil of wire that can spin inside a magnet. When the motor is on, electricity produces a magnetic field in the wire. The coil becomes an electromagnet. The electromagnet spins because its poles are attracted to and repelled by the permanent magnet. This spinning causes the blades of the fan to turn.

 Why does the electromagnet on an electric motor spin?

▶ The fan has an electric motor.

Review

Complete this main idea statement.

1. _____ is produced by the movement of electrons.

Complete these detail statements.

2. An electromagnet is made _____ when wire is coiled around a piece of iron.

3. In an _____ _____, a coil of wire turns inside a permanent magnetic field.

4. In an energy station, a coil of wire turns inside a _____.

Lesson 2

What Are Static and Current Electricity?

VOCABULARY
static electricity
electric current
current electricity
conductor
insulator

Static electricity is the buildup of charges on an object.

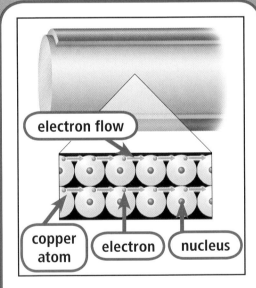

Electric current is the flow of electrons.

In the copper wire, the electrons flow from one atom to another. This produces an electric current.

Current electricity is kinetic energy in an electric current. The bumper cars are powered by current electricity.

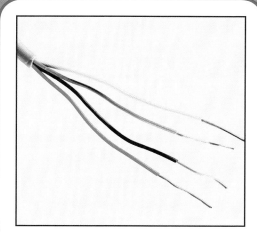

A **conductor** is a material that carries electricity well. Most metals are conductors.

An **insulator** is a material that does not conduct electricity well. Wood, glass, and rubber are insulators.

READING FOCUS SKILL
CAUSE AND EFFECT

A **cause** is what makes something happen. An **effect** is what happens.

Look for the **causes** of static and current electricity. Look for the **effects** of static and current electricity.

Static Electricity

Did you ever touch a doorknob and get a shock? If so, you have felt static electricity. **Static electricity** is the buildup of charges on an object. Most objects have no charge. But when one object rubs up against another, electrons move from one object to the other. The objects become charged. Opposite charges then attract each other.

▼ Lightning is caused by a large buildup of static electricity.

When objects with opposite charges get close, electrons jump from one object to the other. The shock you felt when you touched the doorknob was caused by the electrons moving. The sparks are called *static discharge*. The crackling noise you heard was the sound of the sparks.

Lightning is also static discharge. Negative charges collect at the bottom of clouds. When the charges jump from the cloud to the ground, you see lightning. Lightning is a dangerous static discharge. It can cause a fire and even melt sand into glass.

▲ The cat's fur stands up because the hairs are all charged alike. This causes the hairs to repel each other.

 Explain what causes lightning.

◀ The static electricity on wires discharges a spark.

Current Electricity

Static electricity is potential energy. This potential energy changes to kinetic energy through a static discharge. The kinetic energy of the static discharge can then change into other forms of energy. For example, the electric energy of lightning changes into light, heat, and sound. But static discharges release energy for only a short amount of time. For electricity to be useful, it must be a steady source of energy.

Electricity can provide a steady source of energy. If electrons have a path to follow, they will keep moving instead of building up a static charge. This movement is called an **electric current**. Electricity that flows in this way is kinetic energy. It is called **current electricity**.

To keep the current flowing, you need a constant source of electrons. Batteries provide a flow of electrons for flashlights. Energy stations provide a much larger flow of electrons for whole communities.

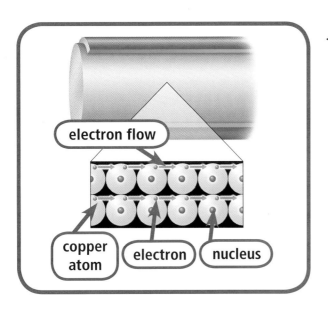

◀ The electrons in the copper wire flow from one atom to another. This produces an electric current.

You use current electricity every day. A lamp is plugged into wires in your home. When you switch the lamp on, electric current moves through the wires. This gives electrons energy.

Different appliances in your home use different amounts of electricity. The amount of electric energy an appliance uses is measured in *watts*. The chart below shows how many watts some appliances use.

 What is caused by electrons moving steadily along a path?

Using Electricity

Some devices use more electric energy than others.

Device	Energy Use
Hair dryer	1600 watts
Microwave oven	1000 watts
Computer and monitor	270 watts
Clothes washer	400 watts
DVD player	25 watts
TV	110 watts
Toaster	900 watts

Conductors and Insulators

Electricity moves more easily through some types of matter than through other types. Electricity moves easily through a **conductor**. Most metals are good conductors. This is because electrons can move easily between metal atoms. So electric current can flow.

Copper is a very good conductor. Copper is used for the wiring in most homes. It is also inside the cords of many electric appliances. But you cannot see the copper wire. It is covered with a layer of plastic.

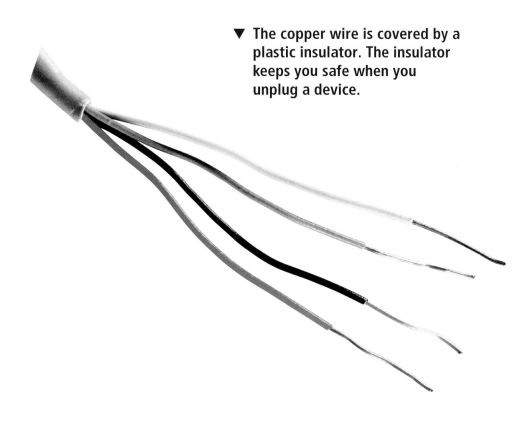

▼ The copper wire is covered by a plastic insulator. The insulator keeps you safe when you unplug a device.

The layer of plastic is an **insulator**. An insulator does not conduct electricity well. This is because electrons do not freely move between atoms of an insulator. Plastic, wood, glass, and rubber are all insulators.

Insulators protect people from electric current. If the plastic around a wire peels off, the wire should be replaced. If you touched the wire carrying the electric current, you could be hurt. The wires carrying electric current also get very warm. If they touch paper or cloth, a fire could start.

 What effect does an insulator have on electric current?

Wood is a good insulator. ▶

Review

Complete these cause and effect statements.

1. The buildup of charges on an object causes _____ _____.

2. If _____ have a path to move in a steady flow, they cause an electric current.

3. When a lamp is turned on, _____ forces electric current through the wire.

4. A _____ carries electricity well because its electrons move _____.

Lesson 3

What Are Electric Circuits?

VOCABULARY
electric circuit
series circuit
parallel circuit

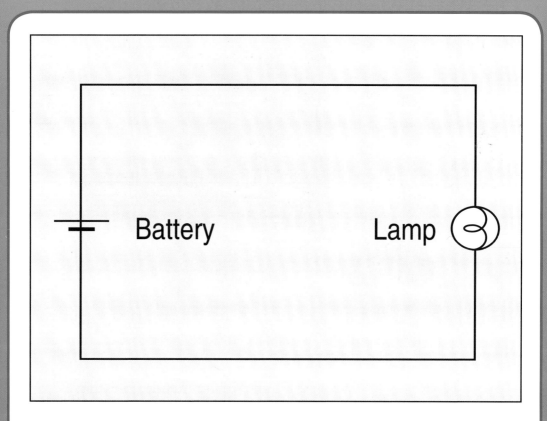

An **electric circuit** is the path that an electric current follows. An electric circuit needs two things for current to flow. First, it needs a source of current. Second, the circuit has to be complete.

In a **series circuit** the current has only one path to follow. Removing any part of the circuit breaks the circuit. The current stops flowing.

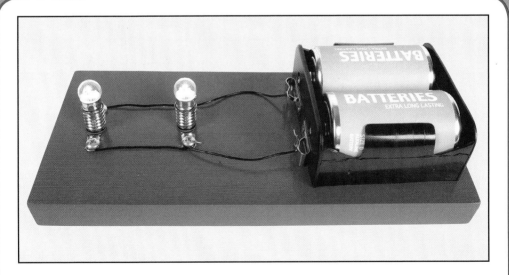

In a **parallel circuit** the current has more than one path to follow. Removing any part of the circuit does not stop the current flow.

READING FOCUS SKILL
MAIN IDEA AND DETAILS

The main idea is what the text is mostly about. Details are pieces of information about the main idea.

Look for information about types of circuits and details about each type of circuit.

Series Circuits

Have you ever tried to turn something on that was not plugged in? The device did not work because its circuit was not complete. An **electric circuit** is the path that an electric current follows. In order to work, an electric circuit must be complete. An electric current also needs a source for the current, such as a battery. A circuit can also have a switch. A switch controls the flow of the current. It does this by opening and closing the circuit. When the switch is on, the circuit is complete. When the switch is off, there is a break in the circuit. A piece of metal inside the switch moves. This stops the flow of current.

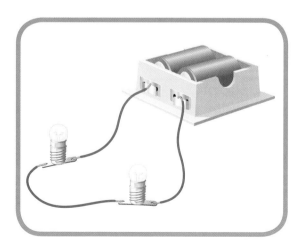

◀ The diagram shows an electric circuit. There are no breaks in the circuit and the circuit has a power source.

▲ Series circuit

One type of electric circuit is a **series circuit**. In a series circuit the current has only one path to follow. All of the parts of the circuit are connected in a single path. If any part of the circuit is removed, the circuit is broken. The current stops flowing.

Series circuits are not used a lot any more. If one part stops working, everything in the circuit stops working.

 What happens to the other light bulbs if one light bulb is unscrewed from a series circuit?

Lights like these used to be in series circuits. ▶

Parallel Circuits

A **parallel circuit** has more than one path for the current to follow. Each part of the circuit has its own path. In a parallel circuit, one part can be turned off, and all of the other parts can stay on. For this reason, a parallel circuit is more useful than a series circuit.

The diagram shows how a parallel circuit can have different devices plugged into it. ▼

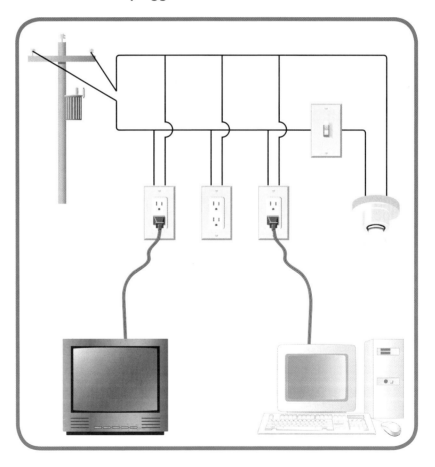

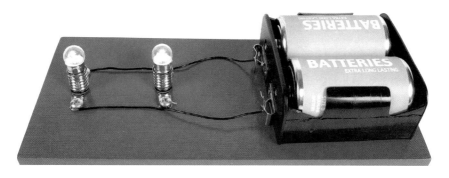

▲ Parallel circuit

Homes, schools, and business use parallel circuits. Each loop in a parallel circuit may have only one switch on it. For example, you may turn on a wall switch and the light on your bedroom ceiling comes on. Likewise, if you turn off that switch and your computer is on, your computer will stay on.

You can connect more appliances to a parallel circuit than to a series circuit. Suppose too many light bulbs are placed on a series circuit. All of the lights become dimmer. In a parallel circuit, adding more lights does not change the brightness of any of the bulbs.

 Name two ways a parallel circuit is different from a series circuit.

◀ The TVs are in a parallel circuit. If one stops working, the others will keep working.

21

Drawing Circuits

Some circuits are very simple. Other circuits are not. When engineers work on a design for a new machine, they do not connect wires to see what works. Instead, they draw a diagram of a circuit. These diagrams use symbols. Each symbol shows a different part of a circuit.

Look at the diagram of the series circuit below. The diagram shows symbols for the battery. It shows the wires and the light bulb. Other diagrams may have more symbols. On the next page is a wiring diagram for a car. It is not simple.

 How are wiring diagrams alike and different?

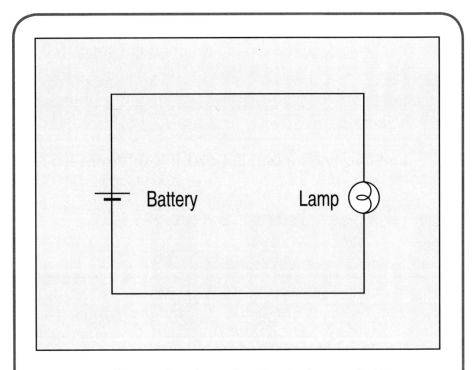

The wiring diagram for the series circuit shows a battery, wire, and the light bulb.

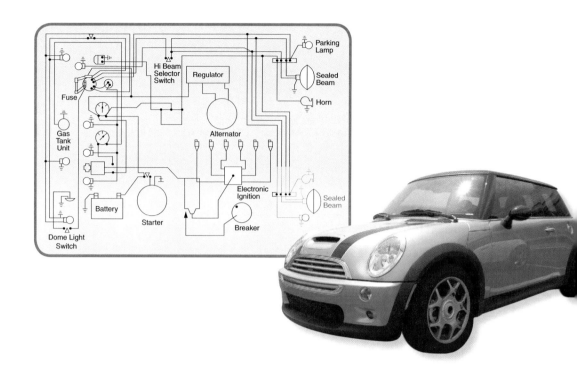

Review

Complete this main idea statement.

1. An _____ _____ must have a source of current and a complete circuit for the current to flow.

Complete these detail statements.

2. On a _____ _____ if one light bulb burns out, all of the other light bulbs go out.

3. On a _____ _____ you can turn on only one device at a time.

4. Engineers use _____ _____ to plan circuits before building them.

GLOSSARY

conductor (kuhn•DUK•ter) a material that carries electricity well.

current electricity (KUR•uhnt ee•lek•TRIS•uh•tee) a kind of kinetic energy that flows as an electric current.

electric circuit (ee•LEK•trik SER•kit) the path an electric current follows.

electric current (ee•LEK•trik KER•uhnt) the flow of electrons.

electricity (ee•lek•TRIS•ih•tee) a form of energy produced by moving electrons.

electromagnet (ee•lek•troh•MAG•nit) a magnet made by coiling a wire around a piece of iron and running electric current through the wire.

insulator (IN•suh•lay•ter) a material that does not conduct electricity well.

parallel circuit (PAR•uh•lel SER•kit) an electric circuit that has more than one path for the current to follow.

series circuit (SIR•eez SER•kit) an electric circuit in which the current has only one path to follow.

static electricity (STAT•ik ee•lek•TRIS•uh•tee) the buildup of charges on an object.